NATURE'S WAY

The Silkworm Story

by Jennifer Coldrey

Photographs by
George Bernard and Dr John Cooke

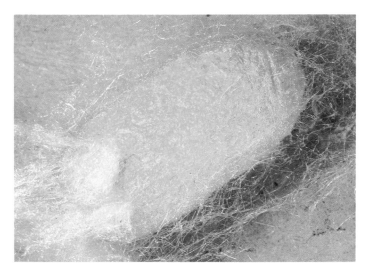

ANDRE DEUTSCH

First Published 1983 by
André Deutsch Ltd., 105 Great Russell Street, London
WC1

Second impression 1986

Text copyright © 1983 Jennifer Coldrey
Photographs copyright © 1983 Oxford Scientific Films
Ltd.

Phototypeset by Tradespools Ltd., Frome, Somerset
Printed in Belgium by Henri Proost & CIE PUBA

ISBN 0 233 97553 5

Grateful acknowledgment is made to J. C. Stevenson
(Animals Animals) for the photographs reproduced on
pages 24 and 25, and to Bob Fredrick for that on page 32
(bottom left).

British Library Cataloguing in Publication Data
Coldrey, Jennifer
  Silk moth.
  1. Silkworms — Juvenile literature
  I. Title
  595.78'1   QL561.B6

The Silkworm Story

The glistening threads of the lovely fabric called silk are made, not by man, but by the caterpillar of an insect – the silkmoth *Bombyx mori*.

This silkmoth came originally from China, where it lived in the wild and laid its eggs on the leaves of the White Mulberry tree. The eggs hatched into caterpillars or silkworms which fed on the leaves and eventually spun a silken cocoon around themselves.

It was over four thousand years ago that the Chinese first discovered how to make silk by cultivating these silkworms and un-winding the silken threads of their cocoons. The art of rearing silkworms to produce silk, called sericulture, was a secret jealously guarded in China for three thousand years, but it was finally uncovered by travellers from Japan, India and Persia, and the knowledge of how to make silk eventually spread across the world. Since then, the native Chinese silkworm, *Bombyx mori*, has been bred and cultivated intensively in many different countries, and its cocoon is now the main source of commercial silk throughout the world. It is no longer found living in the wild and is a totally domesticated animal.

Bombyx mori belongs to the moth family Bombycidae.

Like all butterflies and moths, it has four stages in its life cycle – first the egg, then the caterpillar or larva, next the chrysalis or pupa, and finally the adult moth. What we call the silkworm is the caterpillar stage in the life of the silkmoth.

The adult silkmoth is creamy white in colour, with a thick hairy body, about one inch (2.5 cm) long. Females are slightly larger than males, and both sexes have feathery, branched antennae. Due to centuries of careful breeding and rearing in captivity, the wings have become very small, compared to wild silkmoths, and because their bodies are so heavy, the moths are unable to fly.

The adult moths only live for two or three days, long enough to mate and lay their eggs. They have no mouth and do not eat at all. Female moths produce a special scent from glands at the end of their body and this attracts the males which pick up the scent through sense organs on their antennae. The males dance about excitedly and flutter their wings wildly before mating takes place. Several hours later the female starts to lay her eggs. She lays between three and five hundred tiny, oval, pale yellow eggs, and fastens them one by one onto a mulberry leaf with a special sticky fluid produced from the end of her body. It takes her about two days to lay all her eggs, and then she dies.

In silkworm factories, the eggs are often put into a refrigerator and kept for up to a year before being allowed to hatch. In natural conditions, however, it takes only a few days for the eggs to hatch out into tiny dark caterpillars, about ⅛ inch (0.3 cm) long. The baby caterpillars have shiny black heads and long dark hairs on their bodies. They immediately eat their egg shells and then start to gorge on the mulberry leaves they were born on. They eat ravenously, consuming up to three or four times their own weight during the first twenty-four hours of life. They grow quickly and become very fat until, after four days, their skin starts to split and out wriggles the caterpillar beneath, wearing a new, much brighter coat. The silkworm eats its old skin, then continues feeding on mulberry leaves. It grows fast and soon changes its skin (or moults) again. This happens four times altogether.

Like most other caterpillars, the silkworm's body consists of thirteen sections that fit together like an accordion. At the front of the body are three pairs of legs, each with a single sharp claw, useful for holding onto leaves. Further back there are five pairs of stumpy leg-like limbs, called prolegs. These have rows of tiny curved hooks on them and are useful for clinging to leaves and twigs, as well as supporting the worm's body as it crawls. The caterpillar breathes through a row of small holes on the sides of its body, and its head has powerful jaws which it uses to bite and chew through the leaves of its foodplant.

Older silkworms are a pale cream colour, banded with brown. Their bodies are smooth, almost hairless, and a small horn sticks up from the tail end.

After about four to five weeks, the silkworm is fully grown, being about three inches (7.5 cm) in length, and weighing about ten thousand times as much as when it first hatched from the egg. It stops eating and becomes rather dull and lifeless-looking. It usually crawls away to find a quiet, clean and sheltered spot in which to start spinning. It waves its head restlessly to and fro and then starts to spin the long continuous thread of silk which it weaves around its body to form a golden cocoon.

The silk is produced from two special glands inside the silkworm's body. These open into the mouth through two small holes. The silken fibre, made of a protein called *fibroin*, is secreted into the mouth as two fine liquid filaments. These are immediately joined together and coated with a gummy substance, called *sericin*, which the silkworm produces from two other glands. The resulting single strand of silk is squeezed out of the silkworm's mouth through a small exit tube called a spinneret. The silk is wet and flabby when it first emerges, but soon hardens on contact with the air.

As it spins, the silkworm rotates its head and winds the silk around its body in a figure of eight pattern. The silk is wound from the outside to the inside of the cocoon. The first strands are of very coarse silk — these the silkworm wraps loosely around its body and uses to fasten itself to a twig or other support. As it continues to spin, new loops of a softer, clearer silk are laid down just inside the outer framework, until a close network of glossy fibres gradually builds up around the silkworm's body. After three to five days a soft, egg-shaped case has been formed. When complete, the creature smears the inside with a gummy substance which firmly binds the threads together and thus finishes the work on its cocoon.

Cocoons of *Bombyx mori* are creamy white or golden in colour, and usually about one inch (2.5 cm) long, and ¾ inch (1.9 cm) thick. The silk consists of one long unbroken thread which can be up to one mile (16,000 m) or more in length.

Inside the cocoon the shrunken silkworm changes into a chrysalis or pupa. The old skin is pushed off and a new, hard brown case covers the developing insect. Slowly the pupa changes into an adult moth inside the cocoon. Wings and three pairs of jointed legs begin to develop on the short stubby body, and a pair of feathery antennae start to form on the head. If left to itself, the fully-formed moth would be ready to emerge from the cocoon after about two weeks in the pupal stage. It splits the pupal case and forces its way out of the cocoon by producing a special liquid which dissolves the silken wall at one end.

When silkworms are cultivated, however, this is not allowed to happen, since adult moths would break and damage the delicate silken fibres when forcing their way out of the cocoons. Apart from the cocoons which are kept aside for breeding, most of the silkworms in the factory are killed in the pupal stage, by placing them in hot ovens, or over dry steam, or sometimes by refrigeration. After this, the coarse outer fibres are removed and the cocoons placed in large pans of boiling water. This softens the gum binding the silk together, and loosens the strands, so that the long continuous filament of each cocoon can be wound off onto reels. A single strand of silk from one cocoon would be too thin to use for thread, so the filaments from several cocoons are usually reeled off at the same time and twisted together to form a stronger silk. Later on, in a process called throwing, these raw silk strands are wound onto bobbins, then doubled and twisted together to make a thicker, stronger yarn. The silken yarns are later woven into cloth.

Raw silk thread still contains a lot of gum (sericin) which must be removed at the yarn or fabric stage by boiling in hot soapy water. The silk emerges soft, white and shining. Spun silk is made from short, broken pieces of raw silk, sometimes obtained from damaged cocoons, which are degummed, dried, cleaned and finally spun into a yarn. It is never of such good quality as raw silk, either in its brightness or strength.

Silk has many remarkable properties. Its slender golden filaments are amazingly strong, and also very elastic. It can absorb moisture without feeling damp and it is rarely attacked by mildew. It can also be dyed very easily. Silk is made into sewing and embroidery threads, and is used in many woven and knitted fabrics such as satin, chiffon and velvet. Pure silk cloth is a soft, warm and lustrous material. It is undoubtedly one of the most beautiful natural fibres and is still considered a luxury item today, in spite of the competition from man-made fibres such as nylon, rayon and terylene. Japan, China and Russia are the major world producers of silk, but several Mediterranean countries also manufacture it on a large scale.

Other species of *Bombyx* are also used for silk production, as are some of the wild silkmoths from the family Saturniidae. The Saturniids are the Giant Silkmoths and include some of the largest moths in the world. They are often brightly-coloured with distinctive wing-shapes and patterns, and most species have a conspicuous eye-spot near the centre of each wing. The larvae (silkworms) are also large, usually smooth and plump, and often carry brightly coloured knobs and spines on their bodies. They spin cocoons of tightly-woven silk, often brown, green or silvery-grey in colour. Several species produce valuable commercial silk, although the fibre is coarse and never so fine and beautiful as that of *Bombyx mori*.

The Chinese Oak Silkmoth is one of the most important silk producers of the wild type. It is bred mainly in China and produces tussore or Shantung silk, which is pale buff in colour, and very strong. When spinning its cocoon, this silkworm leaves an opening at one end which it seals with a layer of gum. The emerging moth can break through this without damaging the silken cocoon.

It is possible to rear your own silkworms and watch them spin their wonderful cocoons in your own home. They can be kept quite easily in a jam-jar, tin or plastic box. As the silkworms grow larger, it is wise to put a mesh or muslin cover over the top to allow air in for breathing. Your caterpillars will need continual supplies of fresh mulberry leaves to eat. They can also be fed on lettuce, although they will not then grow as large and healthy, nor produce such good cocoons as silkworms fed on mulberry leaves.

Place a sheet of blotting paper on the bottom of your jar, to absorb excess moisture and prevent the silkworms getting wet and possibly drowning. Droppings should be cleaned out and fresh food provided every day. Never handle the silkworms – they are very delicate and easily damaged. When putting fresh leaves in the cage, leave the old leaves there for a while, until the caterpillars move onto the new leaves by themselves. Remember that silkworms need to stay attached to a leaf for quite a while when moulting their skins. If any worms start to look ill, or become very thin, remove them from the jar, as diseases spread rapidly from one to another.

Keep the jar at normal room temperature, in a quiet corner away from draughts and not in direct sunlight. Silkworms must not be disturbed, especially by noise, when spinning their cocoons. It is a good idea to put one or two twigs in the jar for the silkworms to attach their cocoons to, and for the emerging moths to climb up later. After the cocoons have formed, spray them with water occasionally to prevent the animals drying up inside.

When the moths have emerged you could try unwinding one of your cocoons. First boil the cocoon gently in water for a while, find the loose end of silk, and wrap it round a pencil held horizontally above the cocoon. Start to reel by turning the pencil. It takes a surprisingly long time to unwind the whole cocoon, but remember it took the silkworm four days of continous spinning to lay down its long unbroken thread.

The male silk moth, *Bombyx mori*, uses his plumed antennae to pick up the mating scent of the female.

The scent which attracts the male moth comes from special glands at the end of the female's body.

Silk moths mate by joining the ends of their bodies together. The female's body is fatter than the male's.

The female lays several hundred pale yellow eggs, securing each one to the mulberry leaf with a drop of sticky fluid.

The eggs gradually darken in colour and after about a week the tiny caterpillars start to hatch out.

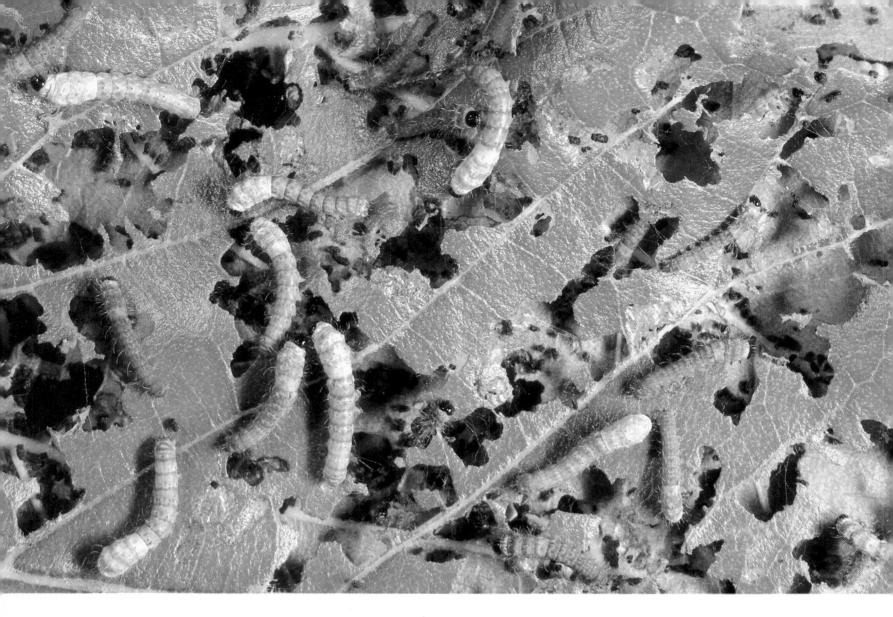

The young silkworms feed hungrily on the mulberry leaves. They grow rapidly, splitting and replacing their skins several times as they develop.

Above The silkworm feeds by gripping the leaf with its three front pairs of legs, while munching continuously with its strong jaws.

Right Older caterpillars are paler in colour and have smooth hairless bodies.

When fully grown the silkworm stops feeding and starts to spin the first silken thread of its cocoon.

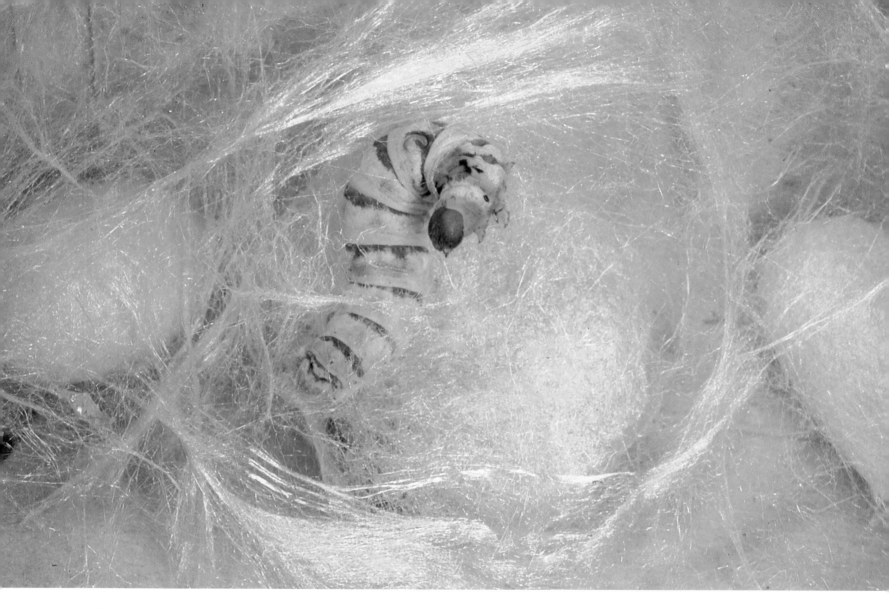

The silkworm spins its cocoon from the outside inwards, making a figure of eight pattern with wide circular movements of its head. The finished cocoons of other silkworms lie nearby.

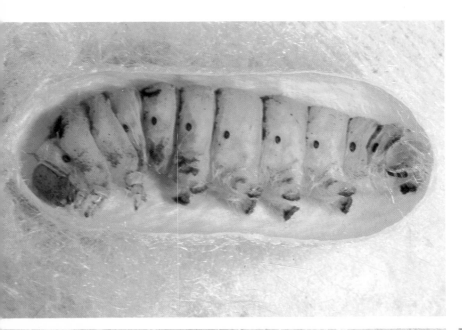

The silkworm, resting inside the finished cocoon, now changes into a pupa, and sheds its skin for the last time.

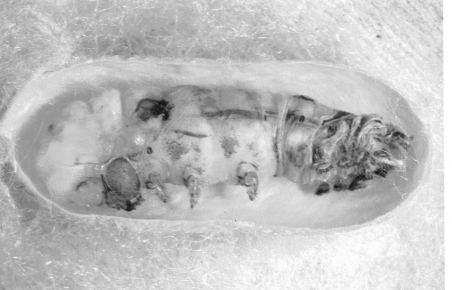

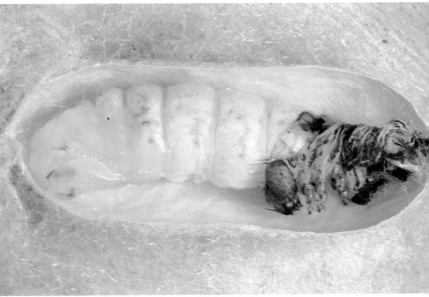

The old skin of the silkworm lies beside the newly formed pupa in its tough brown case.

The wings and antennae of the developing moth can be seen clearly inside the pupa.

17

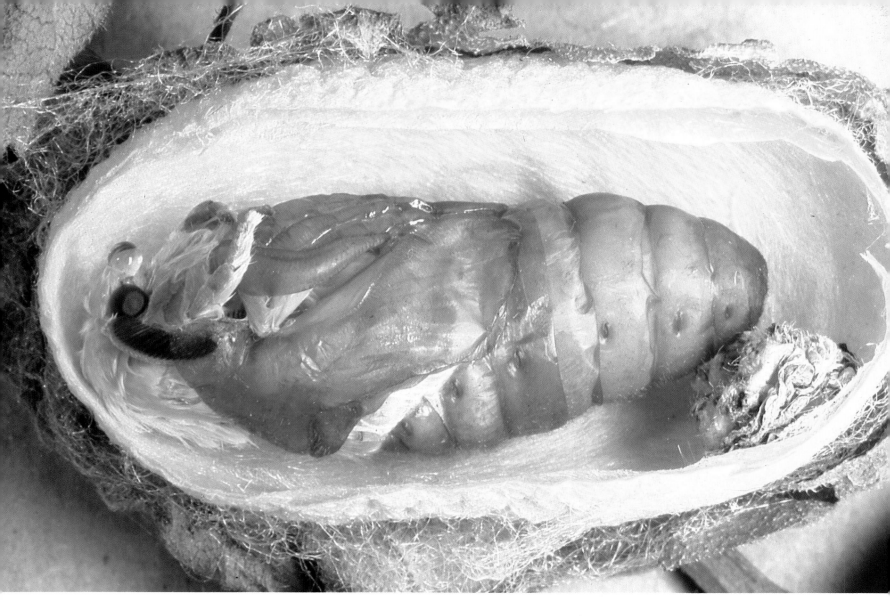

After about two weeks inside the cocoon the silk moth splits the pupal case and begins to push its way out. A drop of fluid produced from its mouth is needed to dissolve away the end wall of the cocoon.

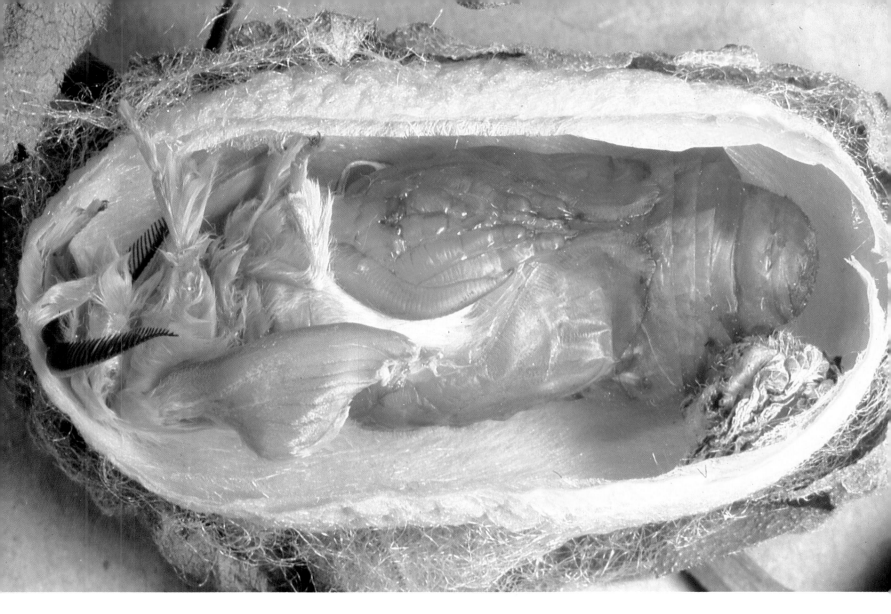

There is very little room for the emerging moth to stretch and move around inside the cocoon.

The silk moth, free of the pupal case, starts to break out through the softened wall of silken fibres.

The emerging silk moth climbs out onto the top of its cocoon.

The newly emerged insect rests for a while to give its damp, crumpled wings time to dry.

As it dries out, its wings become fully expanded. The delicate markings show up clearly on its soft furry body.

In a silk factory in the Far East, cocoons are heated over boiling water to soften the gum made by the silkworm to bind the silken threads of its cocoon. While doing this, the woman in the picture is also reeling off a fine strand of silk.

Strands of brightly-dyed silken yarn are woven into cloth on a loom.

The cocoons of some wild silk moths are also used commercially. The Chinese Oak Silk Moth is one of the most important. Tussore silk is spun from its silvery-grey cocoon.

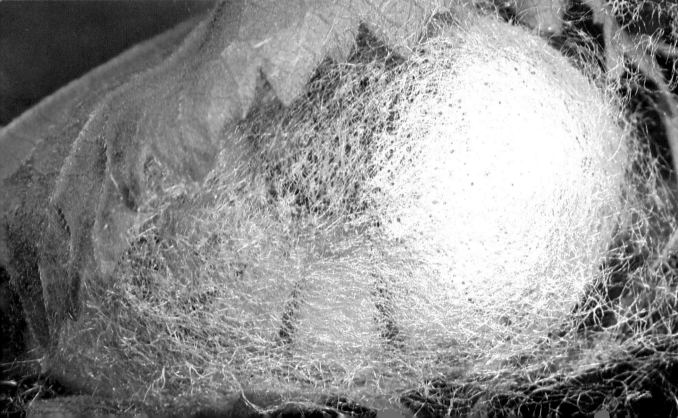

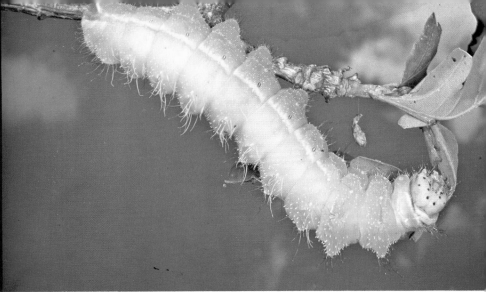

Above The eggs and caterpillar of the Chinese Oak Silk Moth.

Right The male moth has splendid feathery antennae.

The Emperor Moth is the only silk moth living in the wild in Great Britain.

The caterpillar feeds on heather and spins its cocoon among the foliage. The coarse brown silk is never used commercially.

Below The exotic caterpillar of the Emperor Gum Moth, another member of the Silk Moth family, lives in Australia and feeds on the leaves of eucalyptus trees.

Right Before changing into a pupa its gorgeous bright colouring becomes a dark brown.

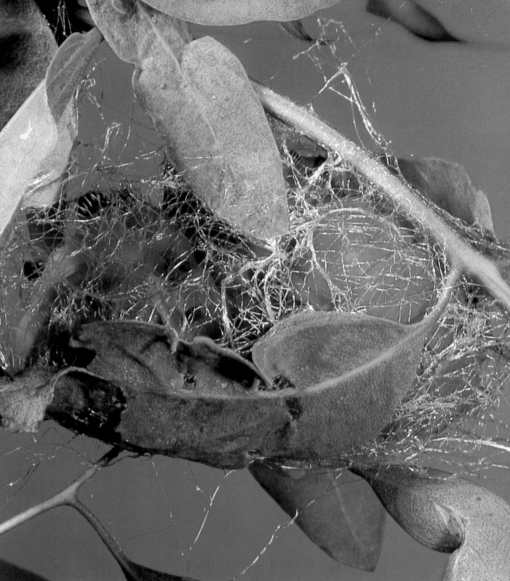

Left During feeding, harmful eucalyptus oils collect in the caterpillar's body and have to be got rid of before it spins its cocoon. *Right* The cocoon is attached to leaves and twigs of the eucalyptus tree. The silk is of no value.

Here are four species of Wild Silk Moth. These include some of the largest and most beautiful moths in the world.

Top left *Rothschildia* from S. America. Top right *Philosamia cynthia* – originates from Asia, now found all over the world. Bottom left *The Moon Moth* – found in India, Japan, and S.E. Asia. Bottom Right *The Robin Moth* – from North America.

PRINTED IN BELGIUM BY
proost
INTERNATIONAL BOOK PRODUCTION